I've Fallen

and

I *<u>Can</u>*

Get Up!

The head to toe guide to why we fall, your risk assessment for falling and a personal guide to fall prevention for a safer and a healthier life.

Alan M. Reznik MD, MBA

FIRST EDITION

ISBN #: 978-1-105-83722-7

NOTICE

The information in this book is the authors' best representation of the top practices and many of the most accepted treatments for common balance problems. Practices vary by physician experience, location, facility and the patient's individual medical condition. The information in this book cannot replace a good physical examination, review of tests and a full understanding of the medical history in a given case. The authors cannot be held responsible for errors or consequences from the use or misuse of the information presented in this book. The authors make no warranties, expressed or implied, with respect to the contents of this publication. The information presented here is not a substitute for advice, opinions or instructions from a physician familiar with the specifics of the patient's condition. It is the sole responsibility of the treating physician, with the full respect of the presenting condition, medical history and experience, to determine the best treatment options for any given patient or condition. Neither the publisher nor the authors assume any responsibility for any injury and/or damage to persons or property.

Table of Contents:

Acknowledgements

First there must be a spark, an idea, a kernel that gets a book going. For this book, that spark has to be the TV commercial "I've fallen and I can't get up!" The title for this book is a direct result of being unable to get that very catchy phrase out of my head for many years. I must therefore give my kudos to the writers of that unforgettable commercial!

Like in the creation of any book, there is a lot of time and effort that goes mostly unrecognized. All the individuals that have given a hand in the assembly of this project and its critical review are so important to the overall completeness, look, feel and educational value. Without them, a book can lose its direction and purpose. I thank Natasha Peavy for collecting and organizing a lion's share of the information and photos for this book and my wife for being a tireless proofreader and sounding board for my farfetched ideas. My dear friend William Schreiber for giving a few kind words after an early read that helped to push me on. Jonathan and Brian (two fantastic physical therapist from our ProPT office) for the balance exercise photos. Once again, my daughter Jane, who helped on my first book "The Knee and Shoulder Handbook for All of Us;" her past efforts made this second book seem very doable. Lastly, my patients who have fallen and I have had the good fortune to help get up again. There is nothing more rewarding than helping someone who has fallen regain their independence.

Photo Credits:

Natasha Peavy

Alan M. Reznik, MD

Jane Y. Reznik

Dreamstime.com

GettyImages.com

Cover: Michele Reznik (Micheleanndesign.com)

On the Floor Looking Up

"An ounce of prevention is worth a pound of cure!" ~Benjamin Franklin

The scene opens with an old woman lying in pain on the floor. She cries out, "Can someone help me? I've fallen, and I can't get up!" Do you remember that commercial? Quickly, the woman pushes a button on her necklace, and a life-saving phone call is made through the magical little device. In the advertisement, these pendants were touted as real lifesavers – a single press of a button and help would arrive at your home. These devices were designed for people with a high risk of falling who had no one around to help them. The idea was to allow elderly people to live safely alone. It became an easy way of providing safety assurance as well as independence. With no one around, Grandma and her relatives need not worry what would happen if she actually fell and "couldn't get up." Better yet, even if she did fall, she would have a simple way to call someone to help her up (or, in many cases, to the hospital) in minutes. The "I've fallen, and I can't get up" commercial was a reality check for hundreds, if not thousands of people at risk for falling. People began to think about what they were most worried about: a life-threatening fall and what they could do if that did happen.

As a third year medical student, I met my first falling victims in the Yale New-Haven Hospital emergency room. Back then, even if we could find out why these patients had fallen, we only addressed the acute problems: we set broken wrists and pinned broken hips. When the problems were not clear, we admitted the victims into the hospital for a few days and tested them, looking for the cause of their unexpected tumble. The causes included dizziness from medical illnesses, loss of muscle control, influence of drugs (prescribed, over-the counter, mistaken doses and illicit drugs too), heart and circulatory problems, high blood pressure, low blood pressure (a sign of dehydration), or alcohol abuse. We provided fluids in the case of dehydration, antibiotics for pneumonia and other supportive care. Once we had solved all the acute problems, the doctors helped the patients get back up on their feet and out of the hospital. Some went home, and some went to skilled nursing care facilities or rehabilitation centers. Most treatments dealt only with the aftermath of the fall. There was very little in the way of intervention for preventing other falls. Even back then, I often wondered if we could do the opposite: prevent the fall, or at the least reduce the risk of a second fall.

Now, as an orthopaedic surgeon (a bone and joint doctor), I have seen many of these unfortunate individuals, young and old, lying on a stretcher in the emergency room with a broken limb. Often at the same time as the initial evaluation, we do our best to figure out the cause of the fall. A fall that is un-witnessed or accompanied by a loss of consciousness often requires a full work-up (that is a medical term for a battery or series of tests and evaluations until we find the answer). Even though we can now recover patients and often learn the cause of the fall, we still do not follow patients for long afterwards to see if there is anything more that we as doctors can do, or what the patients can do for themselves.

So, the question remains: can we stop the falls before they happen? Or, if someone trips and falls, can they learn how to be physically fit enough to bounce back quickly?

This all brings us back to the TV commercial and the little button. These people already knew that they were at risk for falling, and that is exactly why they (or their worried family members) bought the little safety button in the first place. What is worrisome is that whoever needed the button already knew that they wouldn't be able to get up if they fell.

All this talk of falling sounds terrible, doesn't it? No worries – in these pages you will learn why people fall, what happens if you do fall and what can be done to prevent the fall in the first place. You will be more knowledgeable about warning signs of those at risk for falling. We will examine those risk signs of someone who falls, medical conditions that predispose people to falling, how we can correct these problems and the best strategies for fall prevention. We will discuss how to make your home as fall-proof as possible, how to reduce fracture risks in general, what to look for in a loved one at risk for falling and how to best help them stay out of the emergency room and my operating room. Most importantly we will teach you some simple exercises to improve your general health, physical fitness and sport performance all while reducing your risk of falling.

In addition, for those with athletic inclination, we will discuss the special needs of athletic balance and its benefits for performance and injury prevention. Finally, a few pages are devoted to assessing your own risk of falling with our quick self-assessment test.

Wouldn't it be great if there were a new commercial for a different type of button? Instead of an emergency pendant, most falls would have been easily prevented. In this commercial, the person would press the "easy" button and say, “I haven’t fallen and, if I did fall, I **can** get up!!!” After reading this book, it may well happen. So take the first step and look at the check list on the next page. Are you or a loved one at risk? Let’s read on and prevent a fall together!

A Simple Checklist for Increased Risk of Falling

☐ Lack of core strength, limb weakness or muscle injury

☐ Arthritis of any type:

Joint stiffness, deformity, instability or loss of limb motion

☐ Spinal cord compression:

Pinched nerves in the spine:
Herniated disc or spinal canal narrowing (technically known as spinal stenosis)

☐ Loss of muscle control or weakness:

Primary neurologic problems like Charcot Marie Tooth, Multiple Sclerosis, Polio and Muscular Dystrophy, ALS (Lou Gehrig's Disease)

☐ Loss of sensation (most commonly distal neuropathy):

Secondary neurologic problems from diseases like Diabetes, vitamin deficiency, and side effects of chemotherapy or other medications (Vinchristine, Tamoxafen, Cisplatin, etc)

☐ Mechanical problems:

Knee instability: Knee cap (Patella) dislocations, ligament injuries (ACL, PCL, LCL, MCL), cartilage (Meniscus) tears or loose bodies.
Ankle instability: Recurrent ankle sprains, hip weakness, foot drop (usually from neurologic nisorder)

☐ Central balance control:

Middle ear problems, vertigo, Meniere's disease, stroke, brain tumors, AVM (Arteriovenous Malformations), vision loss, cataracts, balance loss, hearing loss, decreased co-ordination or loss of position sense (decreased "Proprioception")

☐ Occupational risks:

Unpredictable environment, crowded work conditions, confined spaces, stairs, ladders, ramps, uneven surfaces or unprotected heights

☐ General conditions associated with aging:

Cardiac, neurologic, osteo-arthritis, balance and strength

☐ Sports:

Players with poor balance, low core strength, poor conditioning, poor field conditions, defective equipment and high risk sports

☐ Unfamiliar environments:

Travel to new location, new home, school or work place

If you or a loved one checked any of these boxes,

read on!

There is important information ahead.

Chapter 1

Are You at Risk for Falling?

Who is at risk for falling – and why?

An environmental and activity based assessment of those at risk

We have all seen babies take their first few steps. It is a very shaky proposition for the baby, but once he or she is up and going, we are all applauding and cheering. Those first few steps are so important. Until a baby can walk, it is up and down and up again with many, many falls in between.

As we grow up, we steady ourselves, and our falls are far less common – until we hit that awkward growth spurt. In adolescence we begin to trip over our own feet as if they are not attached to our legs. Then, with maturity, balance and agility can be taken for granted again.

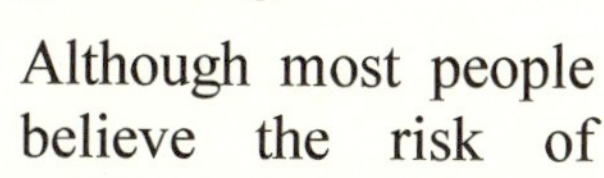

Comstock, Comstock collection, GettyImages

Although most people believe the risk of falling is a problem confined to sick, injured or more mature individuals – it is not. Starting with early childhood, during adolescence, and throughout adult life, the risk of falling is an important problem for everyone. This is even more relevant for those of us that work on uneven ground, unprotected heights and in both recreational and competitive athletics. We all know that at the highest level of sports, even the most skilled athletes also fall.

In athletics, poor balance can reduce sport performance at every level of play. In my office, I frequently attend to young athletes that fall during even the lightest of football tackles, players who cannot hold their ground in a defensive position or throwing athletes that have arm pain as a direct result of lack of hip or leg control. These athletes are in trouble, and the main reason for this problem is the inability to maintain good balance. Athletes also have an increased risk of falling and injury with every play.

Michelangelo Granton, Digital Vision collection, GettyImages

To alter the risk of falling and injury, balance training is a key part of sports preparation – and not only can good balance help prevent unwanted pain, it can also enhance performance. Balance is so important in injury prevention that lessons on how to jump, cut right or left and land properly have been credited with reduction in knee ligament (ACL) injuries in female athletes. Even more interesting is the fact that hip balance has been shown to improve throwing. Not having good hip balance increases the risk of falling, decreases body control in the throwing motion and puts extra strain on the arm. As a result, the *seemingly* unrelated lack of hip control causes an increased incidence of arm injuries in throwing athletes. Many times, a lot of effort is put into treating the arm, shoulder or hand problem. In these cases, even extensive arm therapy, medications and surgery yield no real improvement. Only when the essential *balance* and *leg control* issues are resolved, can the upper extremity problem improve.

Physical Fitness

Physical health and fitness can truly affect your life in both positive and negative ways. Those with poor health have a higher risk of falling. Those with good physical health can still fall, but they are much less likely to endure severe problems afterwards. Different ailments each come with their own hidden hazards – one of which is, of course, falling. Also, if you didn't already know, learn it now:

every part of the body is linked, and if you think your upper body is unrelated to the lower body when you walk, you are wrong! People who have not kept in good physical health usually have weaker muscles and a weaker core. The core muscles are the muscles in the abdomen that help keep the body upright and well balanced while walking or standing. *It's not just your legs that are involved in movement.* The core muscles help control posture, agility and, most important for falling, overall stability. Without stability, how can you stop from falling? Well... it's hard! That's why it is so important to maintain the muscles that will keep you healthy and fall-free for years to come.

Athletics and Sports

Why do athletes playing sports fall? Well, for one thing, they are constantly moving on their feet. Who do you think has a greater risk of falling: someone who works a desk job and sits in a cubicle all day or someone who is constantly on the move? Silly question, isn't it? So, it is no surprise that injury risk in sports is proportional to the hours at play and directly related to the type of sport. Did you know that professional ballet has the same hour-for-hour injury rate as professional football? The types of injury may be very different: there are many more concussions in football (until the recent rule change), but other joint injuries are as common in dance as in football. With the ever-increasing number of hours people play at all types of sports, it is no wonder there are more injuries each year. This is particularly true of women and knee ligament tears. A few years ago, it was thought that males were much more likely to get an ACL tear than females. It turns out the number of hours women participate in sports was lower than men for many years. Now that the sports participation hours for both sexes are more equal, the ACL injury rate in women is four times higher than in men. This is not only true at the high school or college level; we are seeing ACL tears at younger and younger ages. Recently, I operated for the second time in two years on a 13-year-old high-level, ultra-competitive, soccer player with a lot of promise. The first was for one knee (an ACL tear). She returned to full high-level soccer and ended up back in my office with the same injury in the other knee.

Laborers with Occupational Risks

Every single day, billions of people walk places – to their kitchens, their jobs, their car or the grocery store. While walking isn't particularly dangerous, it does take balance – and bad balance can cause falls. Later on, we will cover ways to improve core strength that can greatly improve balance for everyday life. Luckily, most people do not fall simply from loss of balance. But, those who work in certain occupations that are slightly more dangerous than others do. Are you one of them? People with jobs that require work in unpredictable environments, uneven surfaces and unprotected heights are all at a higher risk of falling than the average Joe. A scary but eye-opening fact, stated by the Bureau of Labor Statistics (as noted on http://www.safetycommunity.com), is that falls are **the number one cause of death in the workplace**. In the

Weechia@ms11.uri.com.tw, Flickr collection, GettyImages

workplace, a large part of fall prevention is in understanding more about the "dangerous" or more risky circumstances at work and why they can lead to falling. OSHA, the Occupational Safety and Health Administration, was formed in 1970 to "ensure safe and healthful working conditions for working men and women by setting and enforcing standards and by providing training, outreach, education and assistance." It is this organization that is responsible for many of the work safety rules we see today, and fall prevention is just one area it concentrates on (www.osha.gov).

Common Risks of Falling

Unpredictable or Unstable Environments: What is an unpredictable environment? There are many occupations where the working surfaces move, tilt or are subject to external forces that are truly unpredictable. It might sound odd, but imagine working on a small boat as a fisherman. Balance is a daily, if not a minute-to-minute challenge, so by the nature of a job on water, great balance, and let's not forget a strong stomach, are absolute necessities. Yet, with the best of balance and experience, there are times when the boat will move so quickly that even the most "salty" sailors will get knocked off of their feet. Have you ever seen the show called "The Deadliest Catch"? It is about the men that go out in the cold, insanely rough Alaskan waters to catch king crabs. They also fall in every show. This may be true, albeit to a lesser degree, for all seamen, ferry operators, boat tour staff members as well as others who spend their day out on the water for a living. For the most part, these types of laborers are generally prepared for the daily exposure to motion and their occupation's environment.

Dreamstime.com

On the other hand, a recreational fisherman or passenger on a cruise ship might not be as well prepared to handle such unpredictable environments. They can easily lose their balance and fall. Therefore, overall, it is important to be aware of your surroundings as much as possible – stay alert and keep a look out for incoming waves. Better yet, hold onto a railing as tight as you can like the group above to truly avoid a fall.

Like boats, *any* mode of transportation carries risks for falling. Buses and subways, with their constant stop-and-go, also have unpredictable surfaces for both conductors and riders. On a plane, turbulence can create a very difficult environment for walking down the aisle – so, when the "Fasten Seatbelt" sign is on, be sure to get into your seat as quickly as possible to reduce the risk of falling (or flying inside the cabin) during a turbulent moment!

Uneven Surfaces in the Workplace: Just as moving surfaces create unpredictable environments in work and travel, thereby increasing the risk of falling, an uneven or unsecured surface can do the same. Outdoor occupations in forestry, farming, ground maintenance, lawn and garden care, construction and more all have the risks associated with working on uneven or unstable surfaces. It is easy to imagine increased risks of falling while working in a wooded area or on a pile of rocks. The surface can give from under you or shift with your weight. In these environments, proper shoe wear is very important. In addition, ropes, railings and harnesses may be a necessity for worker fall prevention. Many times, working alone is forbidden in these high-risk environments, again, for safety reasons.

SSteven Puetzer, Photographer's Choice
RF collection, GettyImages

Unprotected Heights: Similar to working on uneven surfaces, working on a pitched roof, cleaning the gutters, painting a house, tree trimming or climbing a telephone pole are all equally risky activities. Again, proper shoe wear is extremely important, and you should not wear loose clothing that could get trapped in a power tool or other equipment. Ropes, railings, safety nets and harnesses are necessary safety precautions. Many times working alone is forbidden for obvious safety reasons.

Confined and Crowded Spaces or Work Conditions: Crowds pose another set of risks. The simple movement of a large group of people in a small space causes collisions, entanglements and falls. If you are at risk for a fall from some other medical conditions like joint stiffness or limb injury (for example a broken leg, cast and crutches), crowds or crowded confined spaces may not be for you. It always surprises me how many people I have just operated on want to go to a baseball, football or hockey game the very next day. Even faced with a large parking lot, stairs, escalators and tight seats, they think it is okay to go - never mind the inability to apply ice to the injury, elevate or protect the sight of the surgery in a stadium seat. Even if I tell them not to, they try anyway. Most of them don't make it too far; the ones that make it into the stadium often have near misses, close calls or occasionally a significant fall. Others come home with a swollen painful limb from not caring for the surgical site properly during the game.

Working in tight spaces also puts people at risk for falls. People who work in tight areas like crawl spaces, ducts, catwalks and tunnels are at risk. People sitting in small control booths for hours frequently get tripped up on wires and equipment when they finally get a break. People that work around large machines in a small space (like engine rooms, generators or printing presses) are constantly climbing or crawling around the machine. Mechanics are at particular risk. They not only have to negotiate the tight spaces, but also the floors in their work areas are often covered with thin layers of oil from years of small spills making the surfaces far more slippery.

These are just a few examples of how our occupational environment puts a worker at risk. The bad news is that these safety hazards exist everywhere, not just in the workplace. Many people working on a more casual basis do not get accustomed to the environments, and they put themselves at increased risk of a fall because of that lack of familiarity. Also, many people "cheat" when it comes to safety precautions by not following rules and all the measures possible to ensure safety. But there is some good news! If you work in an unstable or generally unsafe environment enough, you are actually training your brain to react better to unpredictable circumstances. That training by exposure improves balance and stability in time. The concept of "getting your sea legs" is just that – learning how to balance and not fall while on the unsteady surface of the water. Lastly, for the most risky environments, the Occupational Safety and Health Administration has been developing ways of dramatically decreasing these risks, such as implementing and monitoring specific rules on safety belts and harnesses in difficult workplaces.

Mature Adults

Why is falling more prevalent as we age? When you think of someone who falls frequently, you generally think of an older person. This is because older people are at a huge risk for falling – falls are the *leading cause* of injuries in the elderly in the United States. There are many reasons for this: poor health, lack of physical fitness, exposure to unsafe environments and more. This excerpt from a book on eldercare explains: "Falls are among the most dangerous occurrences for the elderly. A 'fall' can take place in a dark or poorly lit room; exiting a shower, bath, or car; rising from bed or from a chair; bending to retrieve an object from the floor; walking up or down the stairs; and many other places." (Eldercare 911, 54[1]). Had you even thought of all those different scenarios before? Here is a further look into the huge array of why elder people especially are at risk for falling: "Falls often result from: [loss of] balance, eyesight, confusion, disorientation, adverse medication reactions, dizziness, strokes, seizures, or lack of strength in arms or legs." (Eldercare 911, 54). Older people generally have more medical complications than the young. The more mature population (especially women) has a higher rate of osteoporosis and bone fragility. Even if they have a simple fall, they can easily break a wrist, arm, shoulder or hip more easily. Poor muscle strength and lower levels of balance and coordination along with weaker bones explain why they are at an even greater risk of fall and injury.

---

1 Beerman, S., and J. Rappaport–Musson. 2005. Eldercare911. Prometheus Books. For a more detailed / longer list of diseases that lead to a greater risk of falling, see "Head-to-Toe" guide in chapter 8 and the references at the end of this book.

Chapter 2

Are You at Risk for Falling?

Who is at risk for falling – and why?

A *<u>medical</u>* *assessment of those at risk*

Medical conditions can place anyone at risk for a fall. Everything from arthritis to middle ear infections can lead to an unsteady gait and an increased risk of falling. In this chapter, we are going to examine the how and why of the most common and important medical reasons for falling in order to best understand what can be done about each one.

Arthritis, Joint Pain and Instability:

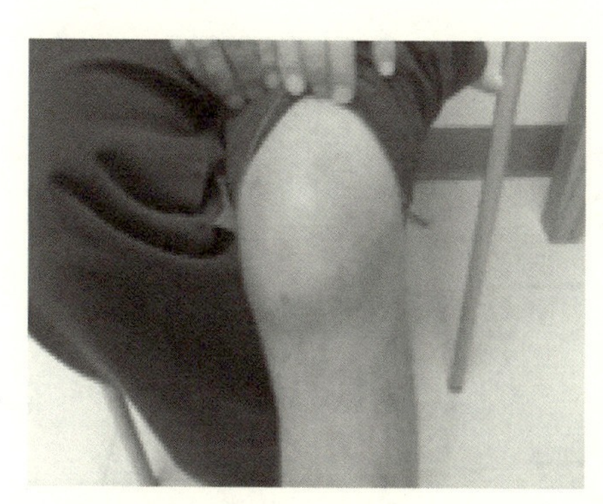

Figure 1: Swollen Knee

If you have mild arthritis or are someone who is plagued with arthritis in many joints, then you know how it feels to have a joint not work properly and limit your activity. And, when your joints are in pain, it's difficult to move them freely. With less mobility and less desire to move, your muscles can weaken, your flexibility lessens, and your general ability to stand and walk with stability is decreased. Even small amounts of knee swelling can also cause the thigh muscles to be weaker. Believe it or not, it has been shown that when small amounts of fluid accumulate in the knee, a reflex turns the quadriceps muscles off. This is one way your body tries to protect the knee and reduce the pain, but this also weakens the leg and increases the risk of falling.

As we talk about falls, we must remember that patients with arthritis have stiff joints and loss of range of motion as well as pain with movement. The combination can be hard to overcome with daily activities, but many people have no choice and push on. With joint stiffness, they have a hard time lifting their feet completely. Therefore, they have trouble with their feet clearing the floor as they move them to walk. As a result, small steps, cracks or uneven surfaces are more difficult for them. Canes and walkers can be required to add stability and help unload the painful joints, and these joints don't react fast to anything. These people are at increased risk of falling. We often see a "Trendelenburg gait" in these people. They may not be complaining of pain in the hip, but they have a characteristic gait (waddle, if you will, from side to side), moving their body toward the painful hip with each step down.

At first the Trendelenburg gait seems to make no sense at all. The weight is centered over the bad hip. How could that make it less painful to walk? To understand this better we need to recall what we learned in the last section on how the hip muscles balance the forces across the hip joint. While walking, the hip muscles add force across the hip joint, and these forces convert tension forces in the bone into compression forces. This is especially true when we walk more quickly, step up a step, jog or run. The body's ability to shift the forces to a compressive force takes advantage of the bone's natural inherent strength in compression. This helps protect the bone from its weaker tension properties. The muscle action protects the hip bone from failing under tension. The side effect of this shift of force increases the total pressure across the hip joint. Therefore, an arthritic hip hurts more when this happens. So what does the body do to drop the pressure across the hip joint and still protect the bone from too much tension force?

For a number of geometric reasons (the internal mechanics of the hip), the tension force can also be significantly reduced by leaning over the involved painful hip as you walk. Yes, leaning the upper body over the involved side actually decreases the forces! Leaning over the painful side sharply reduces the need for the muscles to balance the tension forces. It would seem to make no sense, but it is true.

When leaning over the painful hip while walking, we see movement of the upper body from side to side. It is a very distinctive waddling gait. This waddle is known as a "Trendelenburg gait." The bone loves the force reduction caused by the effects of a Trendelenburg gait. It feels safer to the bone, and the hip becomes less painful. In other words, using a Trendelenburg gait, the body balances the load on the hip bone by shifting its weight over the center of the joint. It works well but, you look a little funny when you walk. You may have less pain, but a skilled eye will see there is something not right with the way you walk.

The Tredelenburg gait itself may not be a reason to fall, but a weak painful hip can be. A painful hip often has a decreased range of motion. The loss of motion may shorten the gait. A shortened gait makes it harder to step over objects in your way. Both of these alter the normal ability to walk. Therefore, a Tredelenburg gait and a shortened stride, both as a result of a painful arthritic hip, also increase anyone's risk of a fall.

Treatment of the underlying arthritic condition should be the first priority. Typically, initial treatment with anti-inflammatory medication helps. If that fails to help, a hip injection with cortisone may improve the symptoms. Surgery to replace the hip may be the last resort, but when done, it often changes the person's life for the better.

Laboratory tests can reveal if your arthritis is related to inflammatory conditions like rheumatoid arthritis, Lupus or Lyme disease as opposed to wear and tear or trauma. Some arthritic conditions are related to other diseases like psoriasis and inflammatory colitis (such as Crohn's Disease). These are clinical diagnoses that can be made knowing the patient's history and physical findings. Conditions like gout or pseudo-gout should also be considered. A consultation with your doctor will be important for your best possible care, especially if pain and warmth are present - even if these sensations come and go. After a good thorough examination, your doctor may want to take X-rays and/or laboratory tests to better make the diagnosis. In all cases, a thorough history, a good physical examination, X-ray evaluation, laboratory tests or testing the joint fluid are the best first steps. (Note: MRI's are often not needed when the plain X-rays show characteristic loss of cartilage, the laboratory tests are positive for an inflammatory process or other specific findings related to known ailments are found on examination or upon reviewing routine X-rays.) Having an accurate diagnosis is the key to finding the best treatment.

Sometimes a visit to an orthopaedic specialist to draw the fluid off and inject the joint (with a local anesthetic and a liquid steroid preparation) is a huge help in reducing the pain and improving the mobility. Looking at the fluid's color and consistency is sometimes enough to make the diagnosis. Further examining or testing the fluid for cells, bacteria and crystals can also help make the most accurate diagnosis.

In general, the treatment is directed at the source of pain and loss of motion in the ailing joint. If the joint is worn out and fails non-surgical treatments, then replacement works best in the majority of cases. A hip or knee replacement in the right patient can be a life saver in more ways than one. It can decrease the risk of falling and improve quality of life by getting rid of the pain. Being pain free allows for increased activity and therefore better general physical condition. Here, again, a painless hip joint helps normalize the gait, increase the stride length and reduce the risks of a future fall.

Nerve or Spinal Cord Compression:

Have you ever had trouble starting your car? Either it makes a lot of noise and nothing happens, or it starts and then shortly stops. "Dead as a door nail" – you cannot move your car, and you may be finding yourself calling a tow truck. Think for a second about how your brain asks each leg to move. There is a thought, you "turn the key," send a signal down the nerves to the muscles and you are off. Well, what if there is a problem with the electrical system? What if the wires are wet, the wires are pinched, the battery is dead or the wires are cut? Similar problems can happen in your body. The nerve may not be conducting the information to the muscles for many reasons. This can cause a lack of control of the legs, weakness, loss of coordination in normal gait or balance issues. This can be more problematic than you may think and some of the most common ways this can happen are explained below.

Spinal Cord Compression (Herniated Disc or Spinal Stenosis):

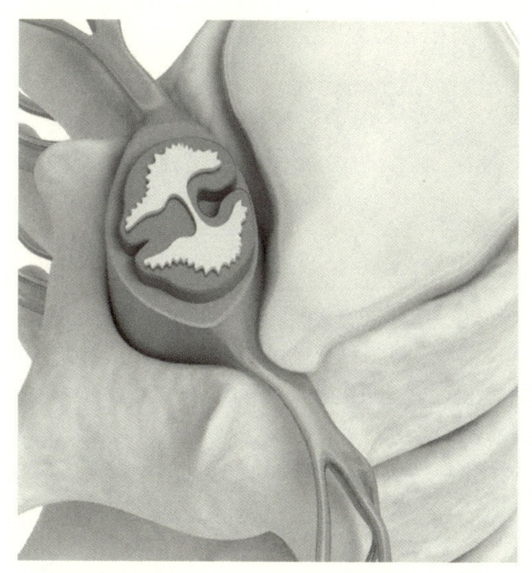

Dreamstime.com

Have you ever heard of a slipped disc? Sciatica? Radiating leg pain? Does your leg or arm go to sleep on you? These are all associated with a pinched nerve in the back, neck or near the hip. How does this occur and why?

Many times, people may have a compressed nerve near where it exits the spinal cord or inside the spinal canal. In the diagram above, we see a nerve exiting the spinal canal and a disc that is leaking (or herniated) and pressing on the nerve. The result is a loss of sensation in the nerve's path and a loss of control of the muscle that the nerve innervates. This can happen in many ways. If the bone presses on the nerve or grows spurs around the nerve, it can have the same effect. In some cases, this is caused by trauma to the spine or a fracture of the vertebrae.

In other cases, it is from arthritis of the spine, in which spurs form and grow. When the spurs are large enough, they can compress the nerves as they exit the spinal canal. In this way, a given nerve or series of nerves lose their ability to carry the full signal to the muscles, and the muscles weaken. Areas controlled by the nerve can also endure partial or complete sensation loss. This can also happen when a disc (the soft material between the vertebrae) leaks out and compresses the nerves. In any of these situations of nerve compression by bone or soft tissue – just like wet wires in your car – the compressed nerves cannot conduct the signal from your brain to the muscle. The muscles don't respond to the abnormal signals to move in a coordinated way, and the ability to control how you walk is just not there.

Imagine a pinched nerve that weakens your ankle flexors so you cannot lift your toes and the forefoot while walking; your toes may catch on the ground with each step. This resultant gait, where a "lazy foot" with dropped toes does not clear the floor, is commonly known as a foot drop gait. A foot drop gait is one cause of frequent tripping and falling. Many patients solve this problem by lifting their leg higher as they move forward. The compensating gait looks like they are stepping over something with every step (a high stepage gait). Sometimes the poorly controlled foot even slaps on the floor with each step. Another tip off that a person may have a foot drop as a cause for frequent falls is the top of the shoe on the involved side is scuffed or worn from rubbing the ground.

Foot drop is a special case in which the nerve that has control over pulling your foot upward is not working well. An undiagnosed foot drop is a major cause of falling. When someone has a foot drop, the foot and toes point down when moving the leg forward, and the toes tend to catch on anything on the ground. If there is nerve compression, then the treatment is to decompress the nerve. If the compression has not been severe for too long, full function can return. Once corrected, the foot drop improves, and the risk of a fall will decrease.

The same might be true for a loss of hip flexion. If your hip flexors are weak, you may have trouble getting your feet to clear a step on the stairs. This would be caused by pinching the nerves at a higher level than a foot drop. This is another troublesome problem that can also lead to falls.

If your quadriceps muscles (the muscles that are in front of your thigh that attach to your kneecap) are weak, you will have difficultly going down stairs, down an incline, squatting or walking downhill. This is caused by the nerve being pinched in between the nerve that causes a foot drop and the one that causes hip flexor weakness.

Many patients have combinations of these issues and making a diagnosis from watching the gait is more difficult. Still, for many patients, a skilled physician can work out a good number of these issues by watching them move around the examining room or watching them walk down the hall of his or her office.

Remember, if you have a foot drop, hip flexor weakness or quadriceps weakness, time can be your enemy. The longer you wait to find the cause, the more difficult and less likely it is that you can gain a full recovery.

There are many other medical causes of nerve related foot drop, hip flexor weakness and quadriceps weakness that are harder to treat than those caused by simple nerve compression. These are discussed later in this book. When they occur, the only treatments that may help are: an ankle brace in the case of foot drop that corrects the drop (something to return the "spring" in your step - pun intended (see the AFO photo)) or surgery to transfer a tendon to the front of the ankle that will lift the foot at the correct time during normal walking. For quadriceps weakness, wearing a knee brace with a drop lock may be required. Hip flexor weakness is much more difficult if not impossible to correct with a mechanical brace. Although braces may help, when possible, the best treatment for nerve entrapment (or compression) causing significant weakness is to decompress the nerve at the site of compression. This means if there is a herniated disc pressing on a nerve root, and it has failed to improve with non-surgical treatments, it should be removed to free the nerve.

Loss of Sensation in Your Feet

(Peripheral or Distal Sensory Neuropathy)

Peripheral or distal (far from the center of your body) neuropathy is the loss of sensation in areas furthest from your spinal cord. Typically, this affects your feet first, then your hands. When you lose sensation in your feet, you cannot feel the ground or floor. With no sensation on the bottom of your feet, little things that we all take for granted become a problem. For example, if you step on a marble, you can feel it. Your brain knows not to put your full weight down, and you adjust your balance. But what if you cannot feel the marble, and you step down with all your weight because you don't know it's there? ...Oops! The marble becomes very dangerous. The same is true if you are on an unstable surface. A small hole, a rock, an uneven floor board and even the edge of a carpet can become hazards to those who cannot feel the subtle change in the surface below their feet.

Simple in office tests for sensation:

Simple dull sharp test

Can you feel a sharp needle tip or the blunt end?
Is it sharp or dull? The doctor randomly tests the skin in different locations to see how your nerves work

Weinstein mono-filament test

(A set of dull tips of different sizes)

Can you feel a tip of a dull rod?
What is the smallest size tip you can feel?

Vibration test

Can you tell when a tuning fork stops vibrating?
The doctor taps the tuning fork and holds the end of it on the skin. You then say when you stop feeling it.
The doctor can then tell how strong the vibration needs to be for you to feel it.

Two point discriminations test

Can you tell when two points are close together?
Can you feel one or two? The closer they are together and you still have the ability to tell the difference between one point and two points, the better your sensation.

People with peripheral or distal sensory neuropathy lose the ability to distinguish sharp and dull feelings on the bottom of their feet. Your doctor can do simple tests for this in the office (see table on the previous page). The patients cannot feel if they have a small cut on the sole of their foot. What's worse, if they get infected or develop an ulcer on the bottom of their feet, they will not know it. This is very important in diabetics since infections are hard to treat and can lead to loss of the skin or limb if not treated appropriately. These people need to check their feet daily for a cut they cannot feel, and they need to see the doctor if their feet are reddened or discolored or have any discharge or drainage.

The Nervous System:

Neurologic conditions that cause gait disturbances:

Diseases:

Diabetes

Autoimmune disorders like rheumatoid arthritis and lupus

Parkinson's Disease

Guillain-Barre syndrome

ALS (Amyotrophic Lateral Sclerosis)

Lou Gehrig's Disease

Charcot Marie Tooth's Disease

Other conditions:

Chronic Alcoholism

Chronic kidney disease

Vitamin deficiency (B6, B12)

Poor blood flow to legs

Underactive thyroid

Raynaud's Syndrome

Side effects of medications:

Chemotherapy drugs:

Cisplatin, Vinchristine, Tamoxafen

High blood pressure drugs:

Amiodarone, Hydralazine, Indapamide and Perhexidine

Drugs that treat infections:

Thalidomide for leprosy, Isoniazid for TB, Metronidazole for bacteria and protozoa, Nitrofurantoin for urinary infections and others

All of the diseases and ailments listed above (gray box) can lead to loss of sensation or an unpleasant tingling (otherwise known as a "pins-and-needles" feeling on the bottom of your feet) – which are not helpful for staying upright. For your risk of falling, it's much more difficult to keep your footing when plagued with a loss of sensation on the soles of your feet, as in peripheral or distal neuropathy.

Treatment can vary greatly in all of the above conditions. Naturally, stopping medications that cause the drug-related sensation loss can help. Changing medications, removing exposure to heavy metals, stopping the use of alcohol, taking B vitamins and controlling diabetes will help. Other situations (more progressive diseases) are more difficult to treat.

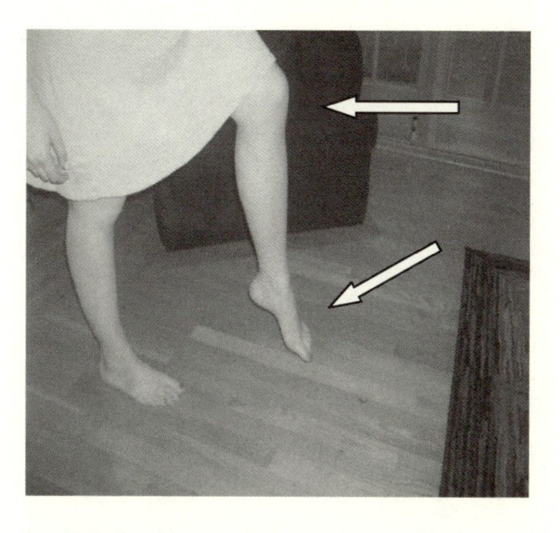

Figure 2: High knee gait to avoid toes catching because of a foot drop

In the cases where a nerve condition causes a foot drop and the foot drop is a result of the loss of motor control, correcting the foot drop with a special brace called an AFO (ankle foot orthosis) can help keep the toes up and reduce tripping over them. The AFO is usually a thin plastic that has an arch support and an extension up the back of the calf to help the foot stay up. Many times the AFO is a springy carbon-fiber material (see figure 3) to add strength and bounce to the gait like a more normal foot. Without the brace, one of the feet lags behind with each step forward. The lagging foot catches anything that is in its way. The AFO holds the foot up and is flexible enough to let the person wearing it push off the ground. In this way, the leg acts like it is working more normally with the foot clearing the floor due to the aid of the lift of the spring action of the AFO. With the foot clearing the floor, the fall risks are decreased.

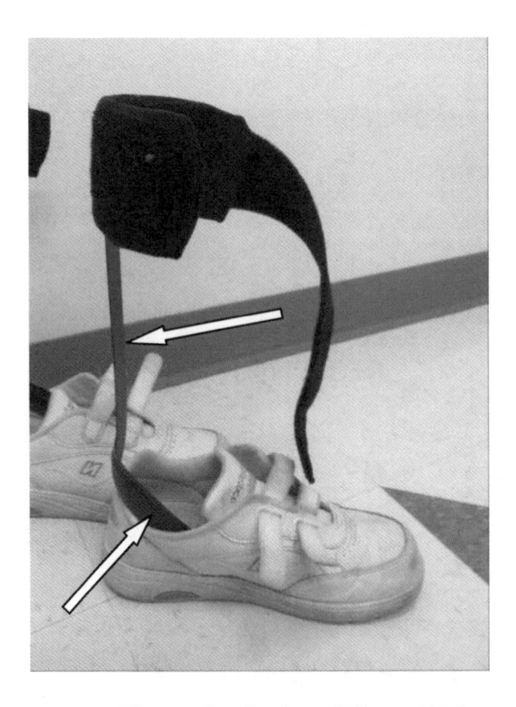

Figure 3: Carbon Fiber AFO

Muscle Control of the Knee

For some people, the control of the thigh muscles has been lost or diminished by nerve damage, trauma to a muscle or a muscular disease. In these cases, the ability to hold up one's own body weight with the knee bend is lost. The ability to go up and down an incline requires that we balance our knees on our kneecaps, which requires quadriceps (thigh) muscle strength. In the photo below, we see a patient using a cane and a brace to "lock" the knee back and use only the hamstrings to support the knee. In this way she can only walk with a hyper-extended knee locked as straight as possible. Any bending while walking and weight bearing will cause a fall.

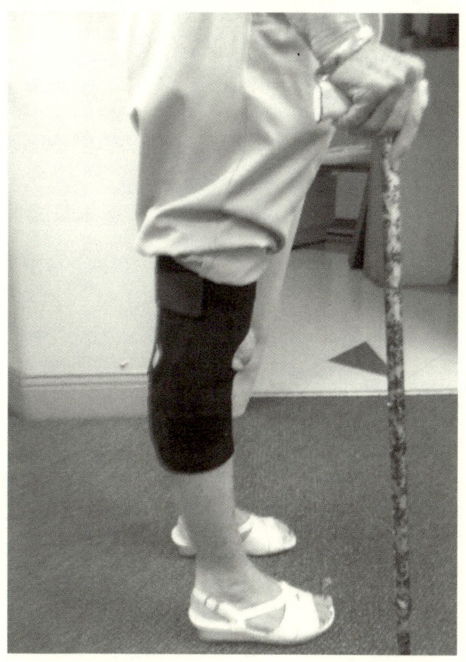

Figure 4: No quads function straight leg gait

Position Sense (Proprioception)

Micheal Fuller, Flickr collection, GettyImages

My patients will ask, “What is position sense? I’ve never heard of it.” It is not easy to explain this strange yet very important sense. However, a few examples from the mechanical world and a self-test of proprioception helps to make it clearer.

Imagine a plane flying on radar or, better yet, a GPS in your car. How does the pilot in a plane or you in your car know where you are? The Radar and Global Positioning System (GPS) allows the pilot and the driver to get information from fixed locations on the ground and the sky. Pinpointing these locations relative to the plane or car creates a frame of reference. With a little math, the plane or car’s navigation system can calculate their exact location. Finding the best air route or quickest way home is now easier than ever before. Well, proprioception is like GPS for your limbs (it occurs to me now that before GPS was common, proprioception was much harder to explain).

How does this work, and why would we need proprioception if our limbs are always attached to our bodies? It seems almost silly to need a "GPS" for our limbs if they never travel away from their attachments. Take the two tests below and let's see the reasoning behind this sense:

First, reach out with your hand, point your index finger straight and touch your nose. It's easy with your eyes open! Now close your eyes and do the same thing. Think about how close you get. Repeat it two or three times, and you will hit your nose with ease even with your eyes closed.

Now get three coins of differing sizes (nickel, dime and quarter), and place them on a flat surface in front of you. Close your eyes again and place a finger on top of each one. Can you tell the difference in the thickness? Pick them up one at a time. Can you feel the thickness difference of each coin? Can you tell which one is which?

Is this a party trick? Not really. Your brain has its own GPS for every joint and part of the body. It knows roughly where your hands and feet are even with your eyes shut. Using its GPS, your brain knows where your nose is and your fingertip, and it can calculate the best path between the two. It also knows how far apart your fingers are, and, pinching them together, it can tell the relative thickness of an object you are holding. Hence you know how thick a coin is, and you can tell them apart. *(By the way, if you are losing your proprioception, you cannot pass tests like touching your finger to your nose with your eyes closed and telling the differences between coin's thicknesses, and you may be the first patient to walk into a doctor's office with the complaint of loss of "position sense" as your primary concern. Your doctor will not believe you since no one has ever complained about that before! You will have to bring this book with you to the appointment to help explain your concerns.)*

The real question is, how does this mysterious loss of position sense affect balance and falling? Think about stepping on uneven ground. How do you know it is uneven? How do you judge how much to press down and at what angle to get over a rock, stick or bump in your way if you cannot feel it? The ability to sense where your feet are at all times without looking is critical to walking on two feet. Without proprioception, a walk in the woods would be almost impossible without looking down all the time.

It follows that when we lose position sense we become more unsure of each step we take. This may be why we sometimes see someone with a cane who just barely touches the ground with it, and they are not limping. It just does not seem like they need it at all. But, if we understand the need for good position sense, we know they are using the cane to help find the ground. They are unsure, even if they don't know it, of where or how far away the ground really is. The cane is just adding more information to help their failing internal GPS system.

Some people believe that knee and ankle sleeves also help in the same way. Some simple braces offer little real mechanical support. They do, however, give a lot of information about the skin rubbing and moving under their surface. The extra sensations of the brace moving on the skin may also help the brain "know" where your affected limb is in space and help the internal "GPS" system do its calculations. In this way, touching the cane to the ground and wearing sleeves, even though they seem to be doing little, may help a lot.

Knee Instability

As you probably know, your legs are a crucial part of standing, and the knee is quite an important part of the leg. That said, if your knee is injured, it is going to be very difficult to stand, balance, hike, work on even

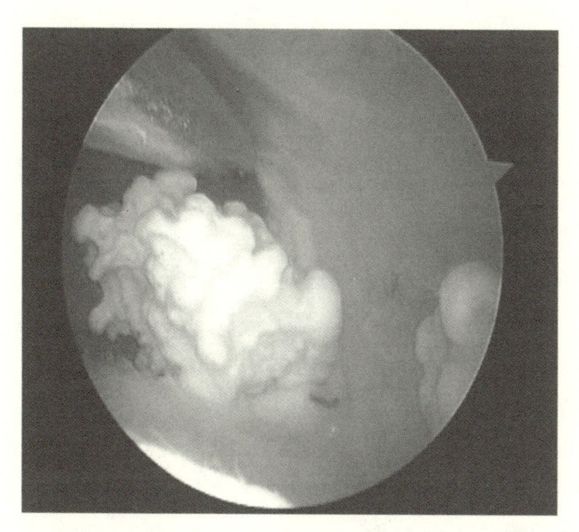

Figure 5: Loose body inside a knee causing locking and giving way (Photo by AMR)

surfaces or play sports. These injuries can include: kneecap dislocations, which prevent full weight bearing on a bent knee; ACL (anterior cruciate ligament) tears, which cause instability with twisting or pivoting on the knee; and MCL (medical collateral and lateral collateral ligaments) tears, which can cause side-to-side instability. PCL (posterior cruciate ligament) tears usually do not cause instability by themselves, but can cause secondary weakness and knee pain over time. Cartilage tears can cause locking of the knee and giving way. Loose bodies can get stuck in the knee and create similar problems.

To treat these issues, you need to address the orthopaedic problem with the knee. More details on all of the knee issues and their treatment are beyond the scope of this book and can be found in my other book:

"The Knee and Should Handbook for All of Us"

by Alan M Reznik, MD and Jane Y Reznik

(see it at: www.lulu.com/spotlight/rezex)

Hips

Many people know that a hip fracture can be a direct result of a fall. Hip fractures themselves are also an indication of many other medical issues. They can be a roadmap into a patient's health or underlying illness. To follow the roadmap and correct the problems can be a key to survival after a hip fracture.

One great example of this is poorly understood and rarely recognized. It is that in some cases, the hip fractures first and then the patient falls. This may be hard to believe, but it is true. Let's see why. Think for a second about bone as a solid material. We know calcium crystals give it its weight and strength in compression, so bone as a material is much closer to cement than steel. Yet, a single long bone like the femur is bowed, and it does bend when loaded (in other words, weight is being put on the leg). When any bone is loaded, the structure sees compression forces on one side of the bone and tension forces on the other. We now know that bone is more like cement than steel; so, how does it handle the tension side? Well, your body is built like it was designed by a mechanical genius. As the bone is loaded in tension, the muscles jump into action and balance the forces by adding compression to the tension side. This drops the tension that the bone "sees" to almost zero. Now, there is a price to pay for this action. The compression forces on the compression side increase quickly. But, as we know, bone is much stronger in compression, so the bone prefers it that way.

Now imagine you are tired after a long day of activity; your muscles are weak. Let's say you also have osteoporosis (your bones have less calcium than they should). Tired, your reaction time is less than normal, and you step down off a step. You slip slightly but you don't fall. Your bone is loaded very quickly with the step, the muscles are late to respond to the stress, and they cannot generate enough force to protect the bone from the tension load. What happens? The bone fails in tension, the neck of the femur cracks, the crack crosses the bone and it starts to fracture. Your hip has broken, and you are still standing on it. The hipbone (also known as the femoral neck) then fails, the fracture is complete and you fall down. This happens very quickly, so most patients cannot tell what happened first; they don't know that they have fallen because their hip was already broken. They assume they fell and then broke their hip. But every once in a while, if you ask someone what happened, they can recall a pain in the hip first (there could be even a few minutes, hours or days of hip pain and difficultly walking) before they fell.

The best **treatment** is really prevention. General conditioning and balance exercises in Chapter 7 are one key to preventing these fracture-falls. Daily walking, good general core strength, good nutrition and calcium supplements are helpful. Knowing your bone density, vitamin D levels and calcium intake are also important.

Pelvis Fractures

Minor pelvic fractures are very common after falls. They can always be related to the fall itself and do not typically need any real surgical intervention. They are frequently non-displaced and can be **treated** with rest and then progressive ambulation with a walker once all medical reasons for the fall are evaluated and cleared. It would be rare to see a pelvic fracture that goes through the socket side of the hip causing a bigger problem and requiring more intervention. But even though it is rare, it does occur.

Sometimes a person has fallen many times but will not say so for a sense of embarrassment or a real fear of loss of independence (as we have discussed in chapter 3). In these cases, the X-ray of the hip and pelvis (note we always take both in cases of a fall and hip or pelvic pain) may show a number of old, healed minor pelvic fractures as well as a new one. Or worse, since these were warnings of falling as a major problem for that person, we can see on X-ray a number of old healed pelvic fractures and now a new hip fracture requiring surgery.

Chest:

In the chest we think of two things: heart and lungs. More accurately, this concerns your circulation and breathing. Heart problems can include high blood pressure, low blood pressure and irregular heartbeats. These all can be a cause of blacking out, fainting and falls. Low blood pressure because of dehydration can do the same. Poor breathing from asthma, chronic lung disease and other pulmonary problems can reduce your ability to get good levels of oxygen in your blood. With low oxygen levels you have poor endurance, muscle weakness and decreased brain function. This can lead to poor balance, muscle weakness, cramps and, of course, falls.

Low blood counts (or anemia) can cause the same problems as low blood pressure and/or lung disease. All three together (heart problems, lung disease and low blood count at the same time) is a bigger problem. Since iron is important in blood's ability to carry oxygen, a low iron level can also have the same effect.

Abdomen:

It is common knowledge that nausea and vomiting cause dehydration. The same is true of prolonged bouts of diarrhea. We already know that dehydration can cause weakness, sensory loss and fainting or falls. Abdominal or GI (gastro-intestinal) upset can also be painful and problematic in and of itself. Worse, GI bleeding can cause anemia (by direct loss of blood). This can be an emergency if the bleeding is acute. A GI bleed is a potentially dangerous cause of an unexplained dizziness, inability to stand up, the inability to walk or a sudden, unprovoked fall. An unexplained fall, dizziness or fainting associated with abdominal pain, nausea, vomiting or diarrhea needs a doctor's attention.

Central Balance Control:

Dimitri Vervitsiotis, Photographer's Choice RFcollection, GettyImages

Without a central control of balance all is lost! Like a space ship going to the moon, we need a navigation system. For it to work well, we need a view of the world around us, gyroscopes and frames of reference. The brain uses the middle ear (our personal gyroscope and the center that controls balance, senses motion, acceleration and deceleration and knows body position), the sense of hearing (like when we step on a piece of glass, and we hear it brake under foot), our position sense of our limbs (remember proprioception?) and the eyes all together in concert to keep us balanced with each and every step. It follows that middle ear problems, strokes, brain tumors, infections and AVM (Arteriovenous Malformations in the brain) all can cause balance issues and falls. And, as we have already learned: balance is paramount in not falling! As we will see in the next section, the middle ear has a very specialized design for this very purpose.

The Ear

The inner ear has a system that enables it to 'know' what your body is doing at all times. Here lie our gyroscopes, accelerometers and position sensors you use each day. To do this, the middle ear has three curved canals, one in each of three plains, positioned inside the ear to sense up-down, right-left and forward-backward motion. Not only does the ear sense motion, but acceleration too – and both help keep our balance.

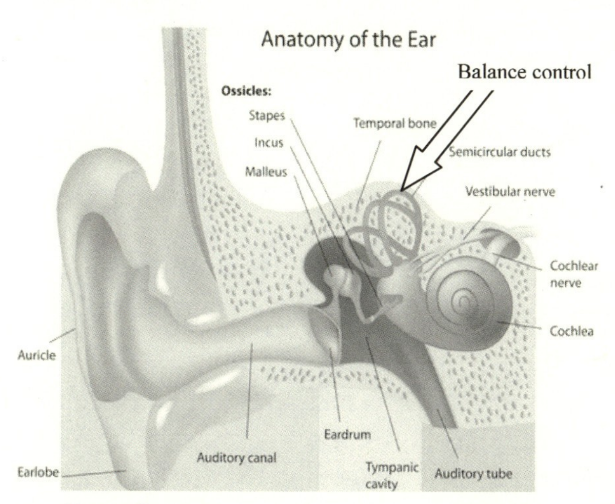

Dreamstime.com

The inner ear system (known as the vestibular system) consists of a group of crystals floating in fluid. There are fine sensory hairs that tell us our position and motion. If the crystals get loose, they hit against the tiny hairs, causing dizziness, nausea or a fall. Dislocated crystals can cause balance disorders, making an individual feel unsteady when walking or even just standing still.

The different systems of the body that work together to sense balance and position include: the vestibular system (which are in the ears, as mentioned above), the visual system (eyes) and proprioception (the body's sense of where it is in space, see the prior section in this chapter). Degeneration or malfunction of any of these systems can lead to balance deficits, and as we already learned, less balance means more falling! For example, with any head trauma, quick head movement from a car accident or a sudden fall can cause crystals in your ear to be knocked out of place. Out of place, the system doesn't work, and walking straight can become a bigger problem.

Treatment for middle ear problems can be as simple as treating an ear infection. If the crystals are dislodged, treatment can also include manipulation of the head to get the crystals back in place. It can be as complicated as balance retraining or medications. If there are chronic infections, surgery for drainage may be needed.

Some middle ear problems and dizziness are much less understood. **Meniere's disease** is one such disorder. It is characterized by recurring dizziness (vertigo), hearing loss, ringing in the ears (tinnitus) and the feeling of pressure in the ear (aural fullness). It often starts with the feeling of fullness and increased ringing and then severe dizziness. An episode may last a few hours and then pass. It can be confused with migraines, stroke, brain tumors, ear infections, multiple sclerosis and heart disease.

Treatment for Meniere's is usually confined to the symptoms, and drugs like Ativert and Valium may help the dizziness. Fluid retention drugs like hydrochlorothiazide may reduce swelling in the middle ear. Balance therapy, hearing aid, steroid and antibiotic injections have all been used to help. Even surgery on the middle ear is used in severe cases.

Other causes for balance issues in the ear need careful evaluation, special studies and focused treatments. Infections can be treated properly when diagnosed, while strokes and heart disease have to be treated at the root cause. Migraines can cause dizziness and have their own set of treatments. Multiple Sclerosis has a number of newer medical treatments and can be diagnosed now more accurately with an MRI. Lastly, brain tumors and AVMs need to be studied with CT scans and/or MRIs and removed when appropriate.

Chapter 3:
What Happens if You Fall?

The costs & results of falling

"Humpty Dumpty sat on a wall,
Humpty Dumpty had a great fall.
All the king's horses and all the king's men
Couldn't put Humpty together again."

You might remember this nursery rhyme from childhood – but you don't actually want to *be* Humpty Dumpty, do you? Of course not! After you read this book, you will be well aware of how to avoid Humpty's dire situation.

So, maybe you have figured out if you are at risk for falling and want to know more, or maybe you're just curious about what happens if you fall. You probably think the results of falling are obvious – it hurts, you become injured and maybe you have to spend money on doctors, prescriptions, surgery and physical therapy. While all that is true, there are other results of falling that you had probably never even thought of. Let's start with the obvious, and then go into the more subtle issues.

Injuries

Injuries in younger people: Younger people are generally more flexible and can bounce back quickly from a fall. But on the flipside, they often feel invincible like superman when it comes to daily life and sports. For a younger person, a fall might not cause a broken bone, but it could certainly cause a twisted ankle. When the immediate result of falling is less severe, they are less likely to be treated and more likely to have multiple injuries to the same body part. Ignoring these injuries until they are truly debilitating is ill-advised – an injury can be much worse if left untreated or more troublesome after repeat injury."

***Injuries in healthy people*:** Sometimes a fall is just an accident, and there was no way to prevent it. Still, when a perfectly fit and healthy person falls, they are still at risk for twisting a joint, pulling a muscle or breaking a bone. A healthy person also might be less in tune with their body than those with chronic illnesses since they are not afflicted each day with a constant reminder of a problem like poor balance or core strength. In fact, a healthy person like you may think that they have good balance or core strength until tested by an unexpected stumble, trip or fall.

***Injuries in the elderly*:** Here is an interesting and important fact to keep in your back pocket: falls are the number one cause of injury in a more mature population. With a little more knowledge and practical advice, you too can avoid *falling* into the category (pardon the pun!) of an injured, elderly citizen. The most common injury from a fall in the more mature population is a wrist fracture. According to Tremblay and Barber from Colorado State University, two thirds of those who fall will fall again within six months. So, a wrist fracture may be an early warning of a hip fracture in the months to come. Sadly, many of these falls are preventable, and one third of falls in the older population are due to environmental issues at home. (see chapter 5 on making your home safer!). **So, it follows that many of the more serious falls have an earlier warning (prior falls and/or fractures) or occur in an unsafe home environment and could**

have been prevented. In general, it is important to know that the risk of falling increases with age. Also, the risks in general are greater for women than for men. It also follows that if we have an increased risk of falling or are at risk because of a medical condition and many falls are preventable, taking the steps to prevent falls can save you from injury. One important thing we can all do is make our homes safer. Falls not only put our health at risk but a fall related injury can also cause you to lose days from work, sports and other activities.

The Costs of Falling

Right after an injury, there is medical care, loss of time from work (for you or the family member that is the primary care giver) and disability. The disability can be both short and long term. These losses in time and productivity have a monetary cost. It's simple math: injuries = lost time + lost $ (income). Both time and monetary losses can be huge results of a fall. In addition to the costs of doctor's visits and surgeries, there are many other costs involved in the aftermath of an injury such as: caregiver costs, medications, home safety equipment and loss of salary due to inability to work.

Loss of work productivity is not limited to the injured person. It can affect a relative that has to take time off work to help an injured family member.

Prevention is everything: As we will learn in this book, creating a safer home environment can help immensely in preventing falls – and, as you will realize, it is also much cheaper than doctor's visits, casts, crutches or a surgery.

Loss of Wages and Productivity

A simple wrist fracture can be treated in a cast or splint for six weeks and then require some therapy to regain motion and strength. A complex fracture may require surgery, bone grafting and a plate. In the end, full use of the hand may take 3-6 months with complete healing at a year. This may not only prevent activities of daily living, but also cause lost days from work, especially if it is the dominant hand or arm. Of course the type of work is also a factor. A clerk that lifts boxes weighing two to three pounds infrequently may be able to return to work much more quickly than an ironworker that does heavy welding at the top of a skyscraper.

The location of the injury also matters; an ankle sprain is more likely to be easier to heal than an ankle fracture, a knee ligament injury, a knee fracture or a hip fracture. However, a computer programmer with the ability to work from home would have less trouble with a broken ankle than a broken hand, and a radio announcer that has to travel from game to game may have less trouble with a fractured collar bone than a fractured knee. (This is because a knee fracture increases risks of blood clots with flying and sitting still for prolonged periods. Whereas a stable clavicle fracture or one that has been plated is far less limiting in that way.)

Loss of Independence

A less obvious result of a fall is loss of independence. Think about it – you fall, and suddenly you are having trouble walking. Or maybe it is so bad that you or your loved one can no longer drive. You may need to be propped up on the couch or bed most of the day, with someone to bring you meals rather than you getting it yourself. Suddenly, daily tasks such as going to the bathroom, running errands or cleaning are much more difficult and sometimes impossible. Imagine trying to eat, showering or going to the bathroom (oh my!) with two broken wrists after a nasty fall. Of course, not all falls result in such extreme conditions, but in many cases they do. When this occurs, no one is happy. It can be bad news all around – bad news for you, your family and your employer.

When older people fall and temporarily lose their independence, they might not want to accept the fact that they are becoming more and more dependent on others. They may be at risk for becoming sick and more prone to diseases and ailments. These are scary thoughts. It is no wonder I often get asked, "When can I drive again?" To some people in the suburbs or rural areas, driving is a daily necessity.

No one wants to accept limits placed on themselves because of their age. They will be unwilling to tell someone in fear of losing their independence. So they might fall, be severely hurt and still experience serious denial. Moreover, they may want to maintain the independence so badly that they may lie about a fall. They may believe they will be forced to move from their home to a "safer" environment when they reveal to others their true state of health or a new physical limitation. We all must assume that no matter how old a person is, they value their own independence, and it is hard for anyone to give it up. For example, if you have an older parent who has been injured, they may not want to tell you, for fear that by helping them, you will begin to think that taking away some of their independence is for their own good. While "helping" a parent may seem like a great idea, it is important to be sensitive to someone who feels like they can no longer take care of him or herself. Understand that they fear being dependent on others.

The fear of being unable to perform simple tasks does not just apply to the elderly – anyone who is injured is now at risk for not being able to carry out everyday tasks that we all take for granted. Think about your day. You wake up, brush your teeth and wash your face. You use the bathroom and wash your hands. You get dressed and then merrily go about your day. Perhaps you make your own meals, go to work or see some friends. Now, think about if you fell and hurt yourself so badly that you were immobilized and could no longer do these things for yourself or by yourself. Imagine you needed someone to help you into your clothes, or you had to rely on someone to prepare all your meals for you – even if you love to cook. This is what it feels like to have your independence stripped away by a bad fall.

Another very real and common cost of a fall is that the injury could cause a person to no longer be able to walk or drive for a specified period of time. After a fall, assisted walking will most likely be possible with the help of a walker, cane or wheelchair. However, even if this is only temporary, it can be a hardship for those that are clinging to their independence. Also after a fall, driving abilities will most likely be suspended for a period of time. It is important to understand the emotions related to the loss of ability to drive and to help the injured person find an alternate means of transportation.

Driving

Chud, Flickr collection, GettyImages

Because most people, young and old (with the exception of those under the age of 16), feel that walking and driving is a huge part of their personal independence, it is important to help the injured party (if they too have lost their ability to drive) find a way to transport themselves through alternate means. This could be: offering to drive the person as much as possible or helping them map-out a subway or bus route.

When there is doubt or concern about someone's ability to drive and personal safety, there are services available to certify their driving skills after an injury. In many locations, an occupational therapist or certified driver rehabilitation specialist can provide testing and evaluations. Car modifications, tools and tips can also help keep an older driver remain independent. A driving evaluation may also help a family take an unsafe driver off the road before it is too late. Safety is always our first concern, and prevention of an accident or an injury to a parent or another party is far better than the regret of letting an impaired person drive and put others at risk.

There are three good websites for more information on this topic:

1) www.helpguide.org/elder/senior_citizen_driving.htm

   To see safe-driving checklist/tips and to help keep driving.

2) www.driver-ed.org

   To find a driver specialist.

3) www.myaota.aota.org/driver_search

   To locate an occupational therapist driving specialist.

Chapter 4

How Can You Prevent a Fall?

"Life is not merely to be alive, but to be well."

~Marcus Valerius Martial

We've learned about who is at risk for falling, and why. We have even discussed what happens after you fall and the costs of falling. We have also touched on a number of medical issues and what to look for when making the diagnosis along with treatments for medical causes of falls. Now comes the practical part of the book: **How can you prevent a fall?**

First and foremost, the short answer is: maintain good physical health! At this point in the book, this category should be a no-brainer. With all the information you have seen, it is clear to see why keeping in good physical health is the number one way to prevent a fall. There are many physical activities one can do to remain in good physical health and, therefore, to prevent falls and other injuries.

Stretching and Muscle Strength

Yoga, pilates and ancient arts like Tai Chi are good exercises for stretching and total-body muscle strengthening. Stretching is important in general for all ages, but sports players should be especially careful to not exercise without stretching first (and after). Continuing stretching and general exercises throughout your lifetime promotes better general health and flexibility. Yoga and pilates are also great for strengthening muscles. Tai Chi is great for balance and strength. The stronger muscles you have, especially the core muscles, the more stable you will be and the less prone to falling (which as you know, can cause unfortunate injuries). Another great benefit to yoga is that it is available for all skill-levels and offered at thousands of locations across the country. These exercises are becoming more and more popular, so jump on the bandwagon and stay fit! Of course, it is possible to stay in shape simply by walking, biking, swimming, jogging and stretching at home. Even something as simple as doing a few stretches while the commercials are playing on the television will make a huge difference in the long run. But don't be fooled by just doing only one thing. I see many patients, especially young athletes, who are in "great shape" and can run miles, yet they have poor balance. I test them in the office. I ask them to stand on one leg and squat. With one or two repetitions, they fall over. They just have very poor hip control and don't know it. They are therefore very prone to injury since they are unable to balance themselves on one leg; they are truly not invincible, and they will get hurt. (See the balance exercises in the next chapter to see what you can do to fix this all too common problem.)

Abdomen and Balance Exercises

Exercises that strengthen the abdomen create a stronger core and therefore lower your likelihood of falling due to balance issues. Pilates, as mentioned above, specifically targets the core. So, if this is your weak spot, rent a DVD or try a class – you will be amazed with the results, and your body will thank you when you are standing taller and falling less!

Walking

Walking is a great exercise and has so many benefits. Walking benefits the cardio-vascular system, coordination, improves mental function and may decrease the overall risk of dementia and even Alzheimer's. Walking trains your spinal cord control of your legs and improves balance especially when walking on uneven surfaces or terrain. Some feel it gives a rhythm to your brain that helps keep it healthy.

If someone you know is at a greater risk of falling, perhaps they simply need a little help. Go ahead and take a walk with someone else who also may need to take a walk. You will be doing them a great service and you may be preventing a fall or two.

Some people have a gait problem or weakness that limits their ability to walk. It may be minor or they may be so unsteady they need a little help to maintain their mobility. There are many assistive devices to help with this – the keyword being "assistive," because the goal is to help the individual remain as independent as possible. Some very useful pieces of walking equipment include: canes, crutches, braces (like the AFO already mentioned or a drop lock brace for those patients with no quadriceps function), walkers (with or without wheels), the Segway, motorized scooters or a huge variety of wheelchairs.

Figure out which is best for you or your loved-one's lifestyle, and get one! It will make the world of difference if they can walk (or simply move)

around with the confidence that they are less likely to fall and injure themselves. Insurance programs treat these items differently and some do not cover the simplest of these items. The more sophisticated items need a review and special conditions before they are covered. Check with your insurance carrier. Many times a diagnosis and a prescription are required before coverage is even considered. You may purchase many of the items in a surgical or homecare supply store.

Non-physical Aspects of Preventing a Fall

Your 5 Senses

Seeing, Hearing, Smelling, Tasting and Touching

You probably take your five senses for granted each day. But they are all extremely important in everyday life. For example, if you cannot see, you are at a much higher risk of falling. If you cannot hear your surroundings, you might not hear a cat meowing at your feet, and you could trip over it. The body was meant to operate perfectly with all of the five senses intact, and if even one is impaired (okay, maybe with the exception of taste!), your risk of falling is increased.

Vitamins and Minerals

It is important to get the proper doses of vitamins and minerals. One very important mineral is calcium. Take supplements if your diet lacks this essential for strong bones! The B vitamins are important for good nerve health and vitamin A helps with night vision.

Awareness and Communication

This is good advice for all areas of life, but it applies to falling, too! Be aware of your own limitations, and if you are responsible for a loved one, observe how they move while looking for signs of trouble. Watch for the signs of poor gait, vision, memory and balance. Do not be afraid to address them – as a parent, sibling, caretaker, or as the injured person yourself! If you feel your body weakening due to an illness or low physical health, do something about it! There are many ways to prevent further injury, and there are many resources available. A simple Internet search or a trip to the library for more information (or the tips and exercises in this book!) can be very helpful. Don't be ignorant: learn, be aware and communicate with your children, loved ones, elder relatives and yourself. Do not be afraid to ask for help, either. Remember: awareness of the warning signs and direct communication is an important component of fall prevention. Prevent a fall *before* it happens – that is simply the easiest and most effective way to keep your family or loved one's health and independence.

Read on to the next chapter for a list on making your home safer!

Chapter 5

Fall Hazards and How to Make our Homes Safer

Most falls occur in your own home, and the reason is simple enough. You spend most of your time in your own home, and you are "too" familiar with your surroundings! What? "Too" familiar?! Yes, the familiarity leads to complacency. Your home is full of falling traps, and don't forget it. Look at your home and the pictures below. How many of you can take pictures like these today without setting them up? I can and did. These photos were taken at a home without prior arrangements. I just walked in and found these hazards in seconds. Look closely at these photos and look in your own home. What do you see? Look at the safety checklist at the end of this chapter and see what you can fix today.

Figure 6: Toys on the floor

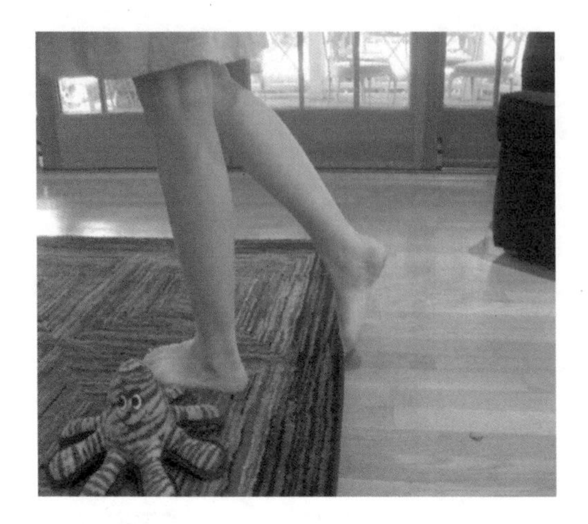

Figure 7: Carpet edges

Figure 8: Exposed, unprotected wire in walking path (Oops! another fall waiting to happen)

Figure 9: Stairs (with a good runner, worn or torn runners are also a potential hazard)

Figure 10: Uneven patio (need I say more?)

Figure 11: Loose garden hose (This beast is in every garden I walked into.)

Figure 12: Improperly supported ladder (open it up, place it on solid ground and have a friend steady it at your side.)

These are only a few examples of common safety hazards in and around the home. Next, we will go into more detail on how to truly safe-proof your living environment.

Safety starts at home, but as we see, so many homes have fall hazards everywhere. Avoid these unnecessary risks by planning ahead and change your home *now* – not after a fall!

Here are some simple and effective ideas:

1. **Rugs:** Replace area rugs with full-room carpet or hardwood. Small rugs are easier to trip on.
2. **Stairways:** Add railings if there are none already, put non-skid runners on stairways and

place safety-gates on both entrances to the stairs. Also, repair broken stairs as soon as they are detected.

3. **Lighting:** Change bulbs to a higher wattage so everything in a room is in plain sight. Add night-lights in the hallways, near stairways and in bedrooms as necessary. Dim lighting can cause people to bump into objects and potentially fall or injure themselves.

4. **Electrical cords:** Remove electrical cords from walking areas – if they cannot be removed entirely (or hidden behind a desk), try taping them to the ground.

5. **Bathtubs:** Put a non-skid bathtub runner at the bottom of the tub, and install a grab-bar to hold on to while bathing. Also place a non-skid bathtub mat outside of the tub, to step on after bathing (the floor often becomes wet and slippery when water and soap splashes out).

6. **Colors:** Paint walls, doors and door frames different, contrasting colors so that an elderly person does not walk into a wall.

7. **Furniture:** Use heavy chairs that do not move, tip or fall when someone sits or stands up.

8. **Backyard:** Wind up hoses, and store them out of walking areas.

Just remember: if you are changing someone else's home, ask first. Explain to them that these changes will help create a safer home, with less risk of accident and injury. Show them what you have done for them; they must be comfortable with the changes.

Chapter 6
What to do After a Fall

You've prepared and prevented, but you've still fallen... now what?

"Happiness lies, first of all, in health."

~George William Curtis, Lotus-Eating

Community is Everything:

Does the person have a community of friends or family that know about their daily activity and whereabouts? If they have fallen at home and cannot get to the phone or the door, will someone know they missed an appointment or dinner out with friends? If the person lives alone, do they have a cell phone or pendant to press to call for help? Will someone know if they are in trouble?

Vial of Life:

"Vial of Life" was created to help in medical emergencies. Rescue personnel often need to act immediately, and there may not be time to stop and gather critical information regarding the patient. Your parent may be too ill to answer questions about his medical history, and family members may not be available. If the emergency occurs in your parent's home, Vial of Life gives rescue personnel immediate access to the information that can help save your parent's life... It's a two-part program usually sponsored by a local fire department, senior service organization, pharmacy, hospital, private community-service-oriented company or other types of organizations. Each sponsor has its own version. Many are called "Vial of Life," but the one in your parent's community may have a different name. Most provide a standard form for your parent to complete, two magnetic stickers that say "Vial of Life Participant" and the 'vial,' a well marked plastic bottle with a cap. "The form will ask for your parent's name and address, the person to call in an emergency and the names and telephone numbers of the physicians involved in your parent's care..." (Eldercare 911, 289-290). I would also add medication and drug and food allergies to the information on the list. Also look into: **Alert1** and **Life Alert** (as seen in the commercial, "I've fallen and I can't get up!).

Medical Care

Of course if you have fallen down and you cannot get up because of an injury to a limb, dizziness or weakness, DON'T GET UP UNTIL MEDICAL HELP ARRIVES! We often see well-meaning friends and family try to get a person up with a broken leg. They cannot walk on it, and displacing the fracture is not only painful, but it can also make the injury worse. Moreover, if someone has low blood pressure and you try to get him or her up too quickly, he or she may pass out again as a result. If a neck injury is a concern, then moving the fallen person can risk spinal cord injury.

The average person is just not prepared for these problems. Try to make the person comfortable where they are. Don't try to straighten a crooked neck! Don't sit someone up who is dizzy. Call 911 and stay with the fallen person. **And don't leave them alone.** While you are waiting, get their medications together, ask what happened, look for reasons for the fall and call the family or closest friends who may be able to help with a good medical history. A history of the medications taken today, including any missed medications, a recent change in medications, reasons for dehydration, nausea, vomiting, diarrhea, febrile illness, seizures, heart conditions, chest pain, dizziness and/or muscle weakness can all be very helpful to the first medical staff on the scene. (Remember to ask if they have a vial of life.) By all means call the victims primary care physician to let them know what happened and the patient is going to the hospital. <u>Knowing if the patient was seen the day before in the office for a new problem can be a huge help to anyone who cares for the fallen person.</u>

Pain Management

Let's face it, even once treated and released from the hospital or after being casted for a fracture, injuries are painful. Luckily, we live in a world where medicine can be prescribed, and pain can be easily treated (for the most part, some soreness is normal and over the counter medications may be enough). Back in the day, maybe it was more common for people to suffer and some never recovered from major illnesses. But today that doesn't always have to be the case. If you fall and are hurt, admit to yourself that you might need some assistance – you do not have to bear all the suffering alone; allow the doctors (especially an orthopedic surgeon if you think something is broken) and those with proper training help you. Do not take your neighbors pain medications. Some pain medications can cause dizziness and falls.

I have to repeat one very important caution here. Some people don't tolerate pain medications well. When this happens, the nausea, constipation or dizziness can be a bigger problem and cause another fall or a return trip to the hospital. Remember, pain medications also have their time and place. They are not candy and can be habit forming. Mostly (except in cancer treatments), pain medications are not meant to be used long term for common problems, aches and pains. If you or someone you know is on long-term pain medications for a seemingly simple problem, you have to wonder if the problem has been treated completely. Resolving the problem causing the pain, even if that means a minor surgery, is often far better than staying on pain medications forever.

Caretakers and Elderly Homes

Can you watch an unsteady person 24/7? In reality the answer is no, it's impossible. No one person can really watch another 24/7. This may be the hardest thing to accept. As well meaning as you are as a concerned family member, there are times when you cannot do it alone. Many times, even though this may mean a loss of some independence for the person being cared for, an outside caregiver is necessary. It may even have the opposite effect by improving the quality of life. A home health aid is one good example. In these cases, a person comes in to help the injured or less able person at fixed intervals. Many times it is for a few hours a day and other times they may sleep at the home to be available. These aids can take the person out of their home as part of their care. Without these aids, many of these people are home bound. Being home bound is not any fun. The aid can really improve the quality of the individual's life.

The same is true of assisted living arrangements. Having meals in a social environment with peers can be liberating to a person who could not get out on their own. If you are stuck at home and cannot walk far or drive, the amount and quality of your daily social interaction drops off. You may be worried about all the details of maintaining the home. With these responsibilities taken over and a good social environment, life at an assisted living apartment can be a great improvement over an individual residence for some people.

Choose the Right Doctor

All doctors have to go through rigorous medical training to become your physician. But some doctors have certain specialties that give them even more insight into a particular area of medicine. For example, an orthopaedic surgeon is going to have expert knowledge of bones and joints and will be able to best help those with a broken bone. Usually, after a fall, the most-often injured part of the body is the hip, so it is wise to seek out a doctor or surgeon specializing in that area, who has performed that type of surgery many times before. For most fractures, a Board Certified Orthopaedic Surgeon is your best bet.

Physical Therapy

In the next chapter, we will take a look at some simple exercises to perform as part of your post-fall physical therapy rehabilitation and other exercises you can do for general strength and balance.

Chapter 7:

Post-fall Physical Therapy & Exercise Guide

Here are some examples of different exercises to perform as part of either recovery from a fall or from surgery – or you can do them each day to practice balance. There are three levels: easy, medium and hard, so jump to whichever section suits you best.

Difficulty Level: Easy or Beginner Exercises

Two-Leg Stance:

Balance for 60 seconds with both feet together:

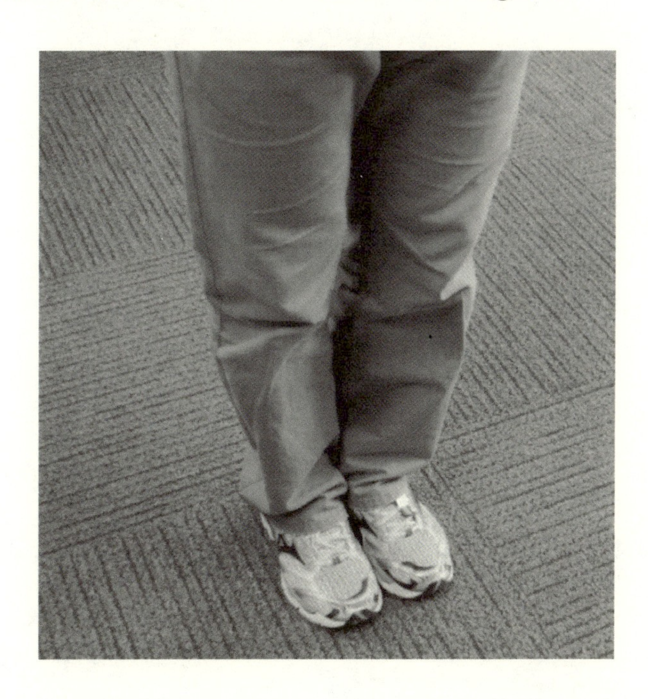

One-Leg Stance:

Balance on one foot, with eyes open, on solid ground (such as cement, linoleum or wood) for 10-30 seconds:

Heel Raise: Raise and lower both heels without supporting your upper body (do not hold on to a table or counter). Hold each repetition for 3-5 seconds; repeat 10 times.

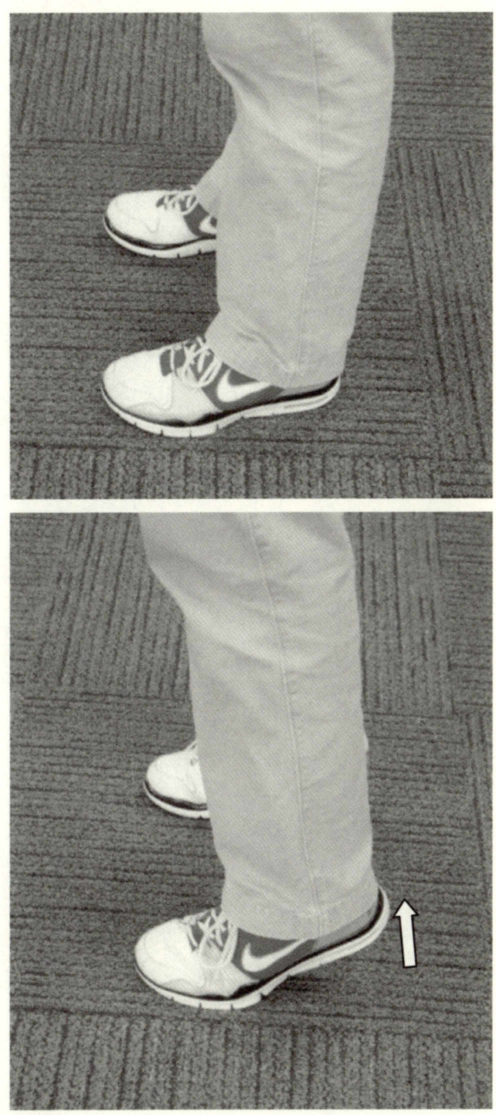

Hip Abduction (Side Kick): Start by standing with both feet on the ground, then slowly kick one leg outward. Repeat two sets of 10 for each leg.

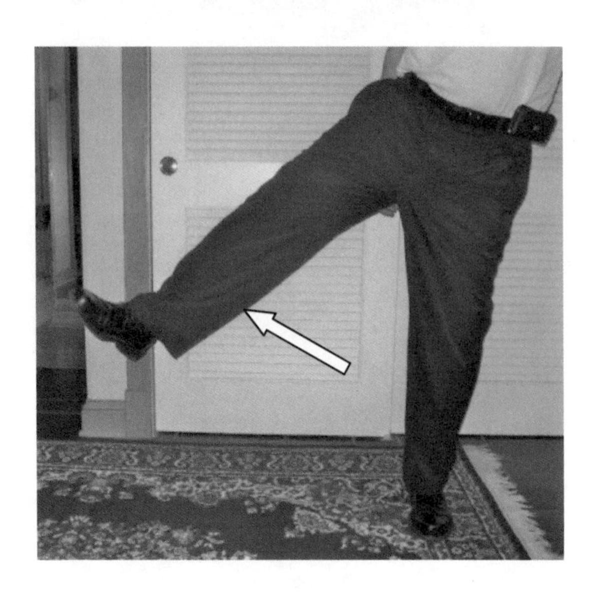

Difficulty Level: Medium

- Slightly more advanced

<u>Semi-Tandem Stance</u>:
Balance with feet slightly separated for 60 seconds:

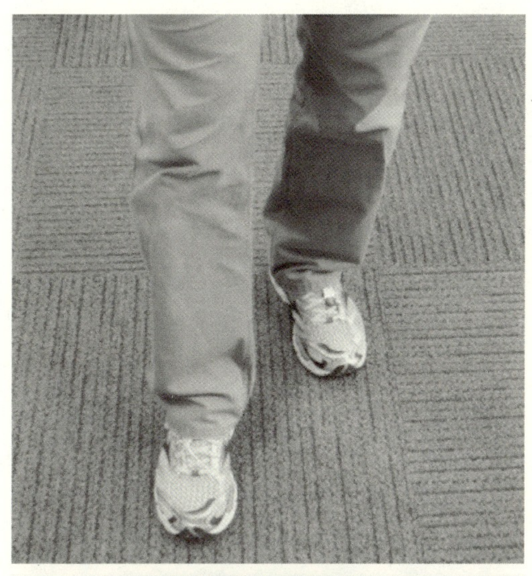

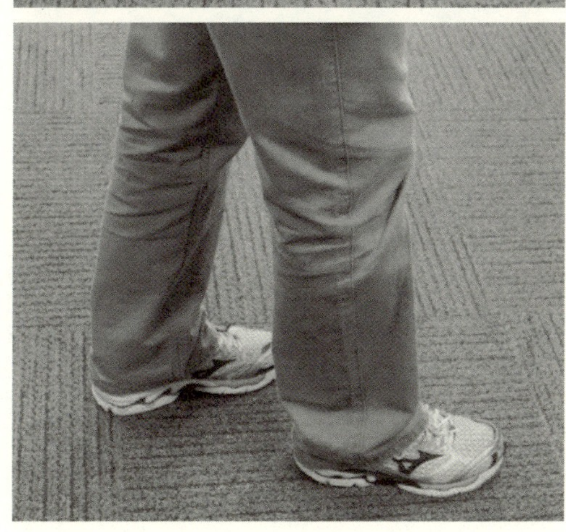

One-Leg Stance: Balance on one foot, with eyes **open**, on uneven ground (such as grass or mulch) for 10-30 seconds:

One-Leg Stance with Ball: Stand on one foot and toss a ball with a friend, or throw against a wall. Do three sets of 10.

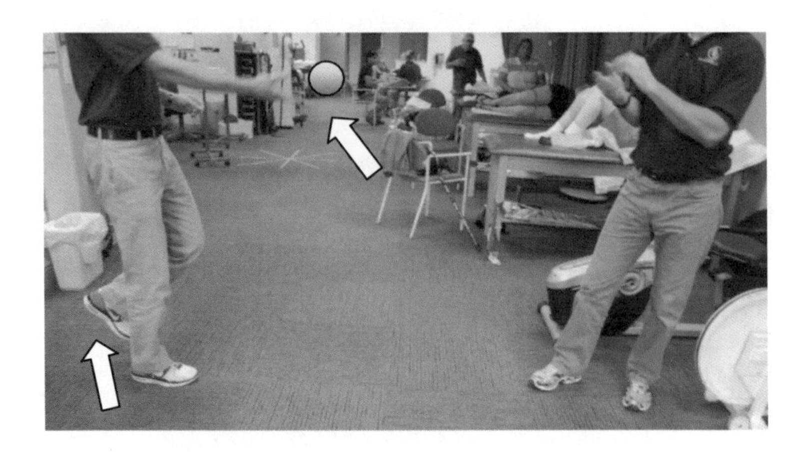

<u>Single-Leg Heel Raise</u>: Without using support, raise and lower your body while balancing on one foot only:

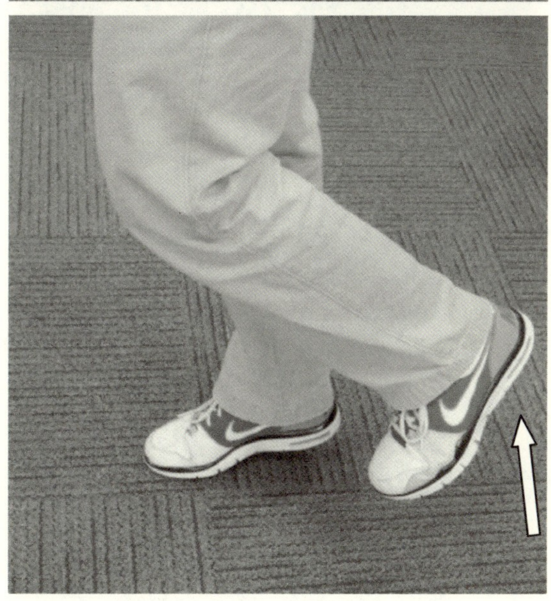

Front Kick: Start by standing with one foot off the ground, then slowly kick the foot forward, and bring back to starting position. Repeat three sets of 10.

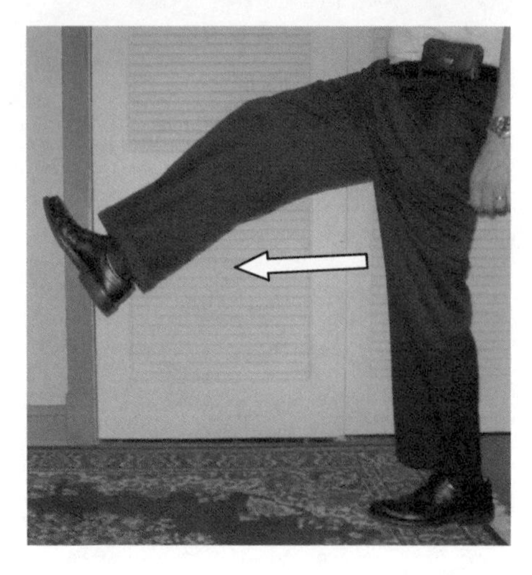

Difficulty Level: Hard

Tandem Stance: Balance with one foot directly in front of the other foot for 60 seconds:

One-Leg Stance: Balance on one foot, with eyes **closed**, on uneven ground (such as grass or mulch) and hold for 30 to 60 seconds:

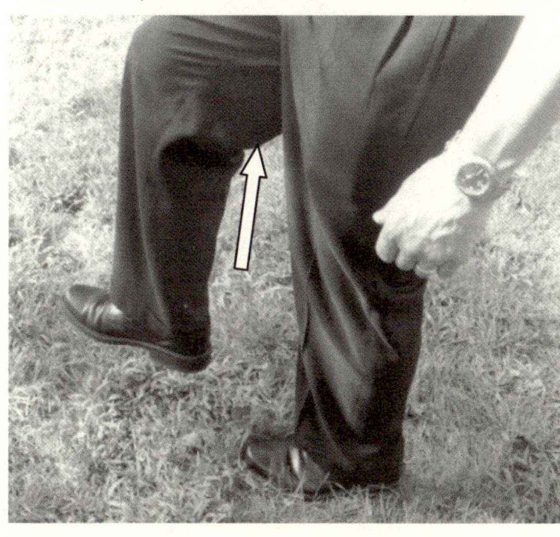

Runner's Pose: Start with one leg slightly raised, then push leg back while swinging arms like a runner. Repeat for 10-30 seconds:

<u>Romanian Dead Lift</u>: First, place a small marker on the ground. Then, standing on one leg, bend over and touch the object, and then stand up again. Do three sets of 5 repetitions:

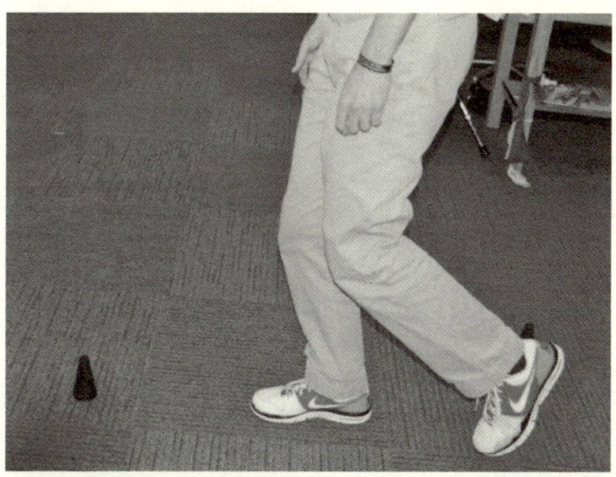

Step 1:

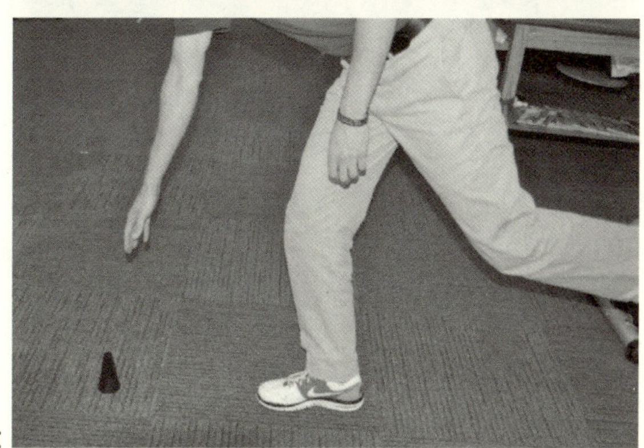

Step 2:

Step 3:

One-Leg Squat: Stand with one foot raised, and then slowly squat and come up. Repeat 10-15 times.

Other Tricks:

Use masking tape to mark out a star like this * on the floor. Then, stand in the middle and tap one foot from left to right, from front to back, or from one diagonal to the other.

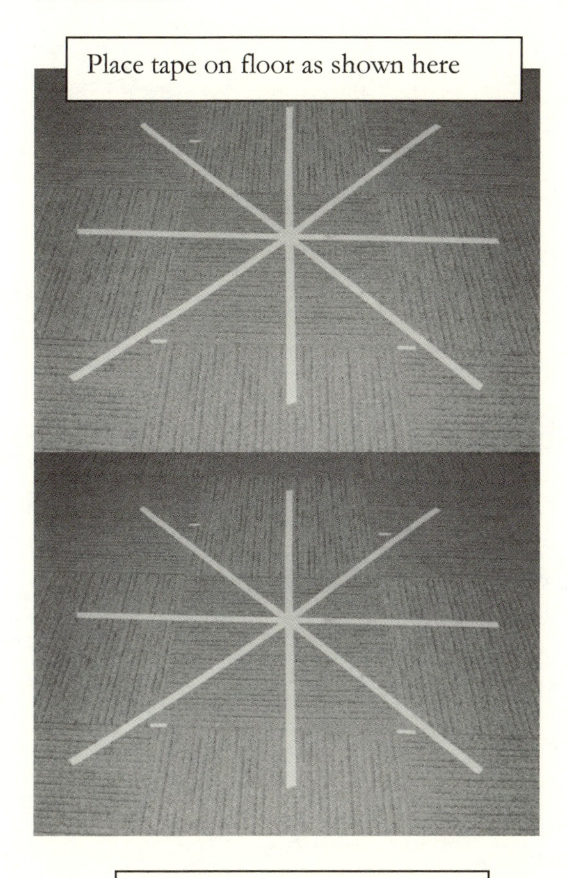

Place tape on floor as shown here

First step on one diagonal

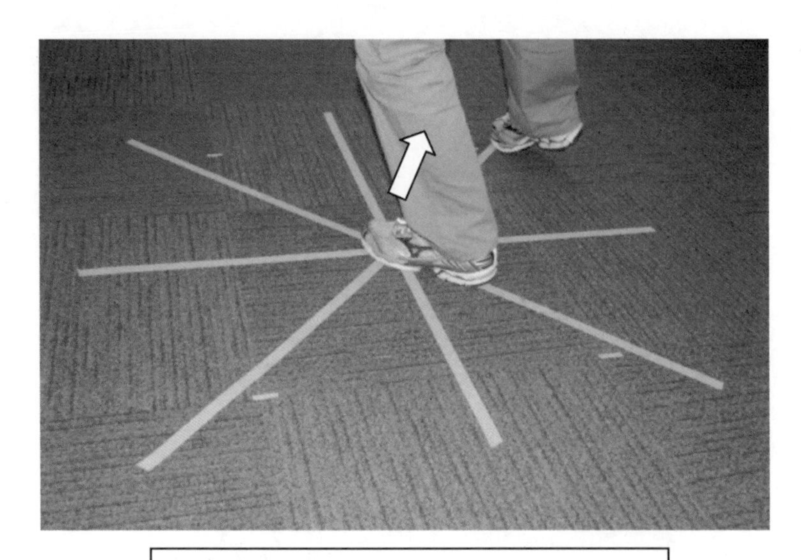

Then move back to center and move to another diagonal

BOSU BALL: This piece of balance equipment is called the BOSU ball because it stands for "**Bo**th **S**ides **U**p."

There are many different ways in which you can use it, but a simple first step is to place it with the flat side down and balance on top. This will be difficult enough at first! Then, as you get more advanced, you can flip it over so that the round side is on the floor. Balancing on this will be much more difficult, but your balance will have improved tenfold!

[images from bosu.com]

One-Leg Wall Catch: This tests balance while using hand eye coordination. This challenges many systems at once. It is great for sports training.

Chapter 8

A "Head-to-Toe" Medical Summary of Why We Fall

The table on the next page is the short but thorough checklist of head to toe reasons for falls. The listed issues are covered in the other chapters in this book. It is set up in a head to toe manner to make it easier to remember all of the components of balance, control and stability. In particular, it would be very helpful in making a good diagnosis when a fall occurs or someone begins to show signs of difficultly with ambulation (walking around). In reading this summary, you can see a snapshot of how all the systems work (or aren't working) together. There are other issues not noted in this table but merit mention. These include the following:

Drug Induced damage to the ear (ototoxicity).

Cervical Vertigo related to issues in the neck and dizziness that is elicited by changing the position of the neck.

Eye issues like ocular albinism (congenital nystagmus) causing the eye to move back and forth constantly.

Ear related pressure sensitivity or sound sensitivity.

Hyperventilation reproduces symptoms, no other exam positives. This can occur in a panic attack.

Neurological Function (Nerve Damage) from a central cause or nerve compression associated with a history of falling or unsteadiness. This could be from stroke, sensory ataxia, migraine, central nystagmus (eye tremor), electrolyte imbalance, loss of blood sugar or brain tumor.

Psychiatric problems with a normal or inconsistent exam or history: agoraphobia, panic attacks, depression or malingering.

Head: Brain (cognitive, vestibular, seizures, MS, hematoma after head trauma, concussion, etc.)
Eyes: Poor vision, loss of depth of field, loss of peripheral vision, cataracts, retina tears, floaters
Ears: Vestibular system, balance control, Meniere's Disease, infections, displaced otoconia
Neck: Spinal cord lesions, discs and spinal canal or nerve compression (canal stenosis), fracture or injury
Chest: Cardiac or heart: hypertension, hypotension, arrhythmias and pulmonary or lung issues (low blood oxygen) or anemia (low blood count), COPD, pulmonary embolism
Abdomen: GI upset, bleeding (causing anemia), nausea and vomiting or diarrhea causing dehydration
Hips: Pelvis fractures (hips sometimes fracture first causing the fall, arthritis in the hips, poor hip strength or control, hip pain, Trendelenburg gait (side to side gait)
Legs: Muscle injury or loss, deformity, arthritis, polio, CP (cerebral palsy), muscular dystrophy and leg weakness from Spinal cord compression or herniated disc
Knees: Mechanical: PF pain MMT, ACL, loose body, arthritis, swelling, etc.
Ankles: Ligament instability, weakness (foot drop) and arthritis
Feet: Loss of control, loss of sensation (diabetic or other Distal Neuropathy), loss of proprioception (position sense)
Toes: Hip flexor weakness or foot drop causing catching of the toes on ground or uneven surface

Chapter 9
Self Fall Risk Assessment Checklist

Do you have one or more of the following concerns:

☐ Cluttered home?

☐ General muscle weakness?

☐ Loss of sensation (diabetic or other)?

☐ Risky occupation (roofer, use ladders or work in unsteady surfaces)?

☐ Poor vision (glasses or cataracts)?

☐ Arthritis or joint pain of any type?

☐ Middle ear problems (e.g. Meniere's)?

☐ Migraines?

☐ Bouts of dehydration?

☐ Seizures?

☐ Spinal cord compression or herniated disc?

☐ Work in confined spaces or unprotected heights?

☐ Breathing or cardiac issues?

☐ Low blood sugar (Hypoglycemia)?

Referenced Websites of Interest

1. www.helpguide.org/elder/senior_citizen_driving - a safe driving checklist/tips to keep driving

2. www.driver-ed.org - find a driver specialist

3. www.myaota.aota.org/driver_search - locate an occupational therapist driving specialist

4. www.osha.gov - occupational Safety and Health Administration website

5. www.safetycommunity.com - safety site

6. www.vialoflife.com - Vial of Life site

7. www.Phillips.lifelinesystems.com - pendant site

8. www.youtube.com/watch?v=AWuKEt96jjs - good video of well thought out balance exercises

9. www.everyday-taichi.com - Tai Chi exercises

10. www.mayoclinic.com/Health/tai-chi/sa00087 - summary of Tai Chi health benefits

11. www.Yogajournal.com - yoga poses, lifestyle and health

12. www.niams.nih.gov/health_info/bone/osteoporis/fracture/prevent_falls.ff.aspw - NIH falling information

13. www.Aspaeris.com - ACL tear prevention shorts and balance program

Selected References:

1. Beerman, S., and J. Rappaport–Musson. 2005. Eldercare911. Prometheus Books.
2. Drachman, D., and C. W. Hart. 1972. Neurology 22:323-334. (Classic article on sorting)
3. Gillespie L. D., M. C. Robertson, W. J. Gillespie, et al. 2009. Interventions for preventing falls in older people living in the community. Cochrane Database Syst Rev. (2):CD007146.
4. Hain T. C., M. Fetter, and D. S. Zee. 1987. Am J Otol 8:36-47. (Head-shaking Nystagmus)
5. Hain T. C. 1997. Approach to the dizzy patient in practical neurology (Ed. J. Biller). This reference gives more detail about the approach outlined here.
6. Halmagyi G. M., and I. S. Curthoys. 1988. Arch Neurol 45:737-740. (Rapid Dolls)
7. Harvey S. A., and D. J. Wood. 1996. The oculocephaic response in the evaluation of the dizzy patient. Laryngoscope 106:6-9.
8. Kroenke K., et al. 1992. Ann Int Med 117:898-904. (Psychiatric)
9. Longridge N. S., and A. I. Mallinson. 1987. The dynamic illegible E (DIE) test: a simple technique for assessing the ability of the vestibuloocular reflex to overcome vestibular pathology. J Otolaryngol 16:97-103. Acta Otol (Stockh) 103:273-279. Otolarygol HNS 92:671-677.
10. Lord S. R., S. T. Smith, and J. C. Menant. 2010. Vision and falls in older people: risk factors and intervention strategies. Clin Geriatr Med 26:569–581.
11. Lord S. R., et al. 2008. Effective exercise for the prevention of falls: a systematic review and meta-analysis. J Am Geriatr Soc 56:2234–2243.
12. Madlon-Kay D. J. 1985. J. Family Practice 21:109-113. (ER)
13. Muir S. W., K. Berg, B. Chesworth, and M. Speechley. 2008. Use of the Berg Balance Scale for predicting multiple falls in community-dwelling elderly people: a prospective study. Phys Ther 88:449–459.
14. Nedzelski, et al. 1986. Otolaryngol 15:101-104. (Otology setting)
15. Nelson J. R. 1969. Neurology 19:577. (Neurology setting)
16. Stevens J. Falls Among Older Adults: An Overview. Accessed March 24, 2010.
17. Studenski S., S. Perera, K. Patel, et al. 2011. Gait speed and survival in older adults. JAMA 305:50–58.